PUMA

Horst Scheibert

Schiffer Military/Aviation History
Atglen, PA

Opposite page: A factory-new Puma (Sd. Kfz. 234/2). Its form is impressive.

PHOTO CREDITS
Federal Archives, Koblenz (BA)
Combat Troop School 2
Schröder Archives
Frank Archives
Podzun Archives
Scheibert Archives

Translated from the German by Dr. Edward Force.

Copyright © 1994 by Schiffer Publishing Ltd.

All rights reserved. No part of this work may be reproduced or used in any forms or by any means – graphic, electronic or mechanical, including photocopying or information storage and retrieval systems – without written permission from the copyright holder.

Printed in the United States of America.
ISBN: 0-88740-680-7

This book was originally published under the title,
PUMA,
by Podzun-Pallas Verlag.

We are interested in hearing from authors with book ideas on related topics.

Published by Schiffer Publishing Ltd.
77 Lower Valley Road
Atglen, PA 19310
Please write for a free catalog.
This book may be purchased from the publisher.
Please include $2.95 postage.
Try your bookstore first.

A medium armored observation vehicle (Saurer) of Armored Artillery Regiment 119 of the 11th Panzer Division. Note the official symbol of the division above the artillery symbol at right on the nose, and the additional symbol (a ghost) of this division on the left side of the nose.

The Heavy Armored Reconnaissance Vehicles of the ARK Series

Since the heavy (8-wheel) armored reconnaissance vehicles of the GS Series, developed in peacetime, proved to be not very usable because of their sensitivity to shot and dust, the Tatra firm was given a contract in 1940 to develop a new, lower and less sensitive armored reconnaissance vehicle.

By the end of 1941 the first two test models were turned over to the Army Weapons Office. Instead of the previously used box frame, they featured a self-supporting hull, a 220 HP Diesel motor, a higher fording ability, better armor, new ventilation and larger, wider tires. The chassis (type designation ARK) and the body were made by the Büssing AG.

Many requests for modifications – especially to the powerplant in view of its use in North Africa – and the necessary testing delayed the beginning of series production. Only in mid-1943 could the first new heavy armored reconnaissance vehicles be delivered to the troops.

There were four versions:
– Heavy Armored Reconnaissance Vehicle (2 cm – Sd. Kfz. 234/1)
– Heavy Armored Reconnaissance Vehicle (5 cm – Sd. Kfz. 234/2)
– Heavy Armored Reconnaissance Vehicle (7.5 cm L/24 – Sd. Kfz. 234/3), and
– Heavy Armored Reconnaissance Vehicle (7.5 cm L/48 Pak 40 – Sd. Kfz. 234/4).

In addition, plans existed for:
– Heavy Armored Reconnaissance Vehicle (2 cm anti-aircraft)
– Heavy Armored Reconnaissance Vehicle (1.5 cm or 2 cm triple)
– Heavy Armored Reconnaissance Vehicle (7.5 cm KwK 40 – Ak 7 B84)

As to the numbers of heavy armored reconnaissance vehicles of the ARK series that were built, there are very varying statistics. Presumably, though, there were only about 1000 of them. Unfortunately, there are very few photographs.

A heavy armored reconnaissance vehicle with a 7.5 cm KwK L/24 (Sd. Kfz. 234/3). In 1944 there were 88 of these produced. It was to be found in the armored reconnaissance units as a "support vehicle."

A Heavy Armored Reconnaissance Vehicle (2 cm), Sd. Kfz. 234/1.

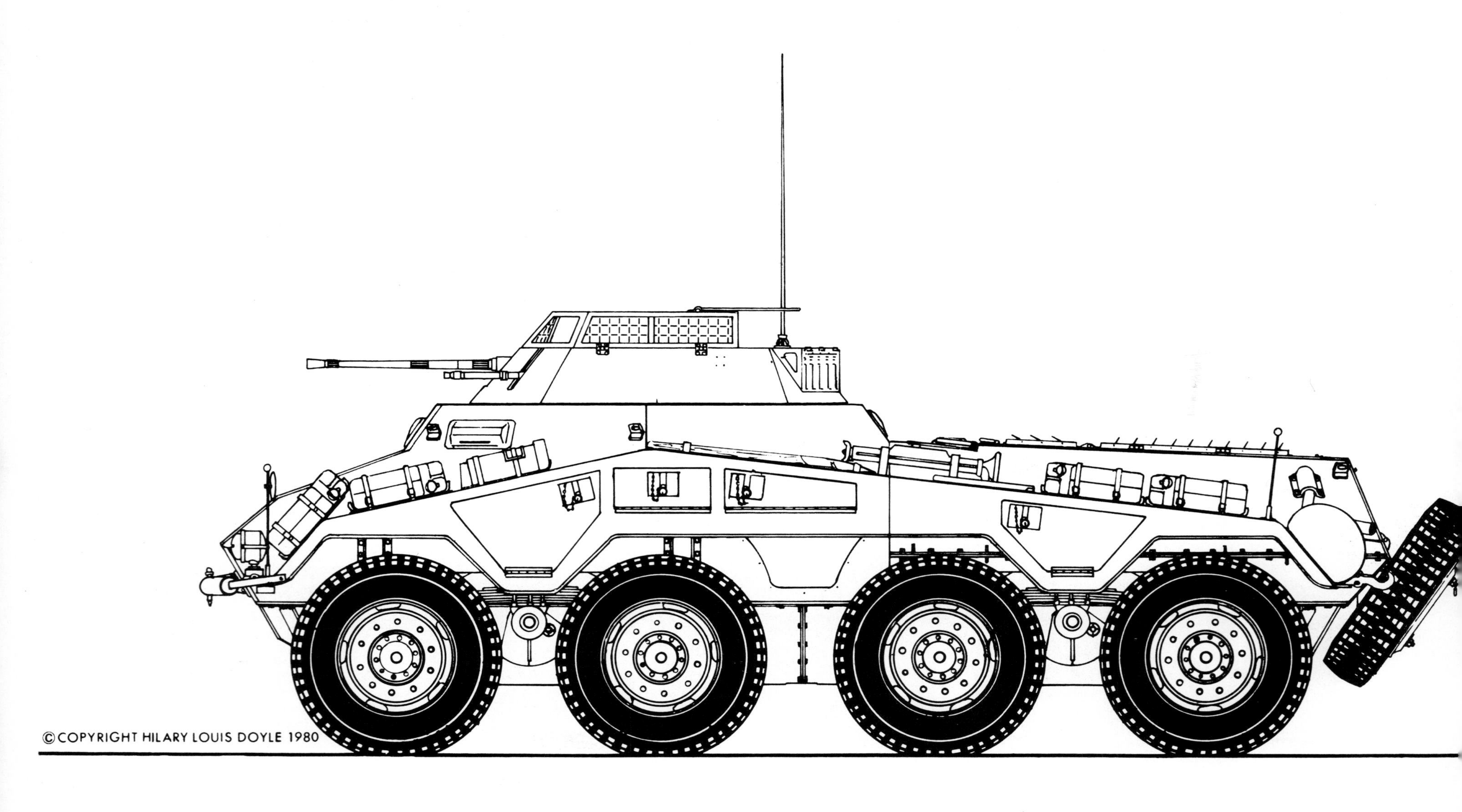

Heavy Armored Reconnaissance Vehicle – Sd. Kfz. 234/1

It was the basic vehicle of the new ARK Series, and was intended to replace the older Sd. Kfz. 231 and 232 of the GS Series. It had a low open turret, made by Daimler-Benz (Berlin) with the collaboration of the Schichau Works in Elbing. The turret was fitted with a hanging mount that was equipped with one 2 cm KwK 38 and one MG 42 (coaxial).

Like all the models in the ARK Series, it had only a staff or umbrella antenna (the latter instead of the oddly shaped bow antennas of the older series). With its Diesel engine and a large fuel tank, it was able to expand its radius of operation enormously from that of the earlier 8-wheel armored reconnaissance vehicle, to 900 kilometers. The crew consisted of four men. In all, several hundred were produced. At the war's end, there were plans to equip it with a 2 cm anti-aircraft gun and a 1.5 cm or 2 cm triple set. These were not put into production.

As opposed to the 231 – its ancestor – it had an open turret, similar to that of the 4-wheel armored reconnaissance vehicle, Sd. Kfz. 222.

Upper left: While its predecessor was driven by a carbureted engine, it had an air-cooled 12-cylinder Diesel engine producing 210 HP.

Above: At the right front is a 234/1. It has an opened grid over its open-topped turret. The wheel covers, typical of all heavy armored reconnaissance vehicles of the ARK Series, can also be seen.

Left: This view from above allows a look into the turret, set at "5:00."

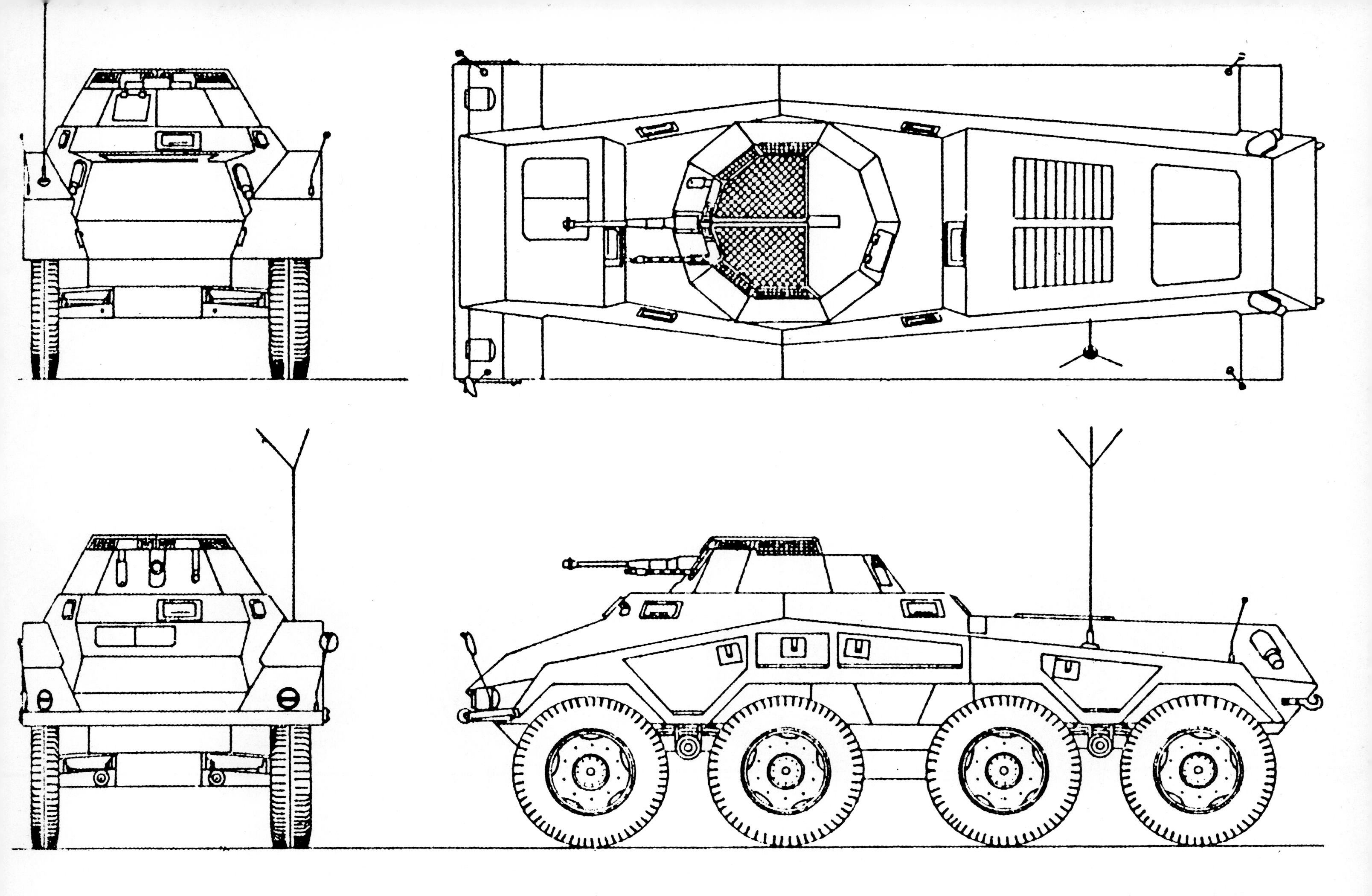

This four-way drawing shows – as is often found, unfortunately –a ten-sided turret. There was only a six-sided turret, as the top view on the opposite page shows.

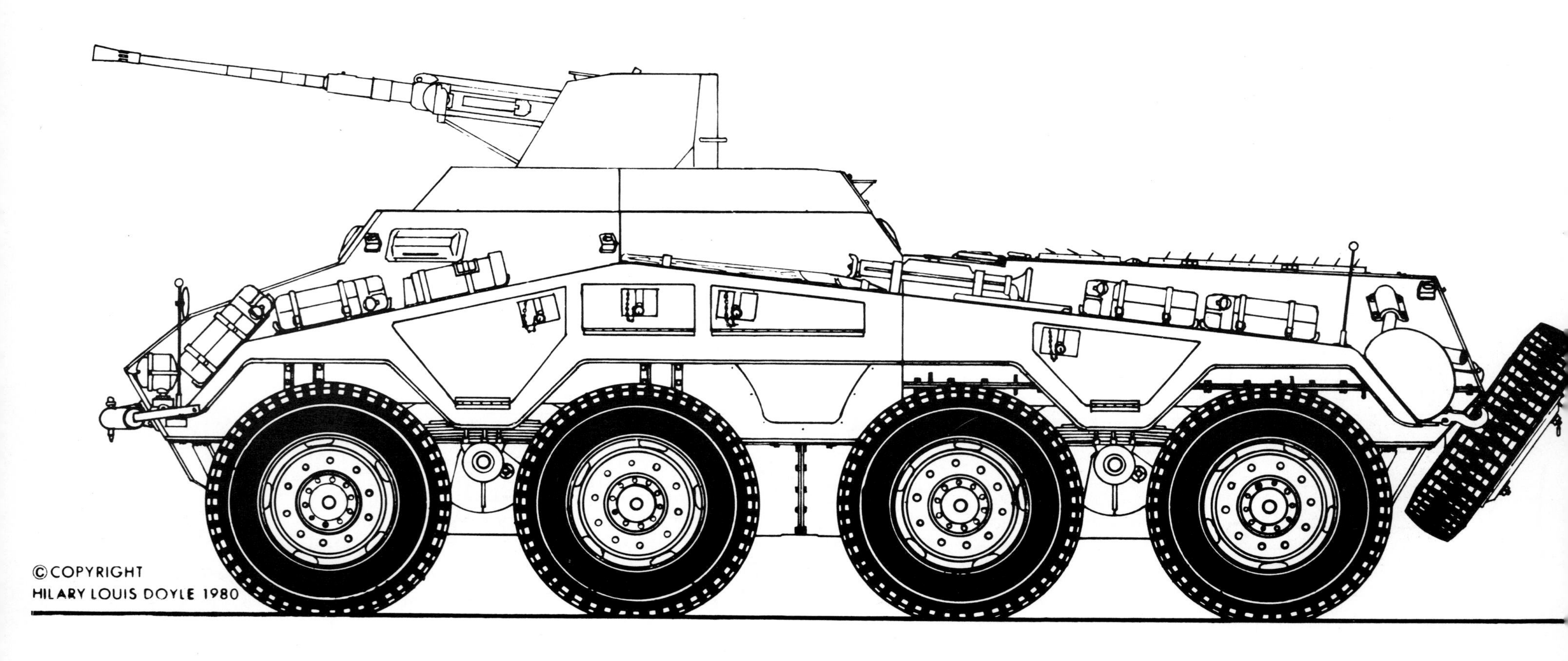

This heavy armored reconnaissance vehicle with a 3.7 cm anti-aircraft gun in a protective shield was only a plan that was never put into production.

Heavy Armored Reconnaissance Vehicle – Sd. Kfz. 234/2 "PUMA"

With its rotating closed turret it looks somewhat different from the rest of the ARK Series, which otherwise only had open fighting compartments. It was armed with a 5 cm L/60 – also used in the Kampfpanzer III beginning with Ausf. J. The inferior steadiness of a wheeled vehicle as opposed to a tracked one, though, required the installation of a muzzle brake, so as to mitigate the recoil. In addition to the tank gun, it had a coaxial MG 42 installed.

Its crew consisted of four men. Its weight was 11.74 tons. It had six forward and six reverse speeds, could reach 80 kph, and carried 55 rounds of ammunition. The armor of its "pig's-head" (Saukopf) gun mount was up to 10 cm thick. It was designated "Puma", and 101 of them were built.

Because of its good-looking form along with the modern technology of the ARK Series, it is still an admirable vehicle today.

Below: The most spectacular heavy armored reconnaissance vehicle of the ARK Series was the Sd. Kfz. 234/2. Unlike all the other vehicles in this series, its turret was closed at the top. It was given the name "Puma."

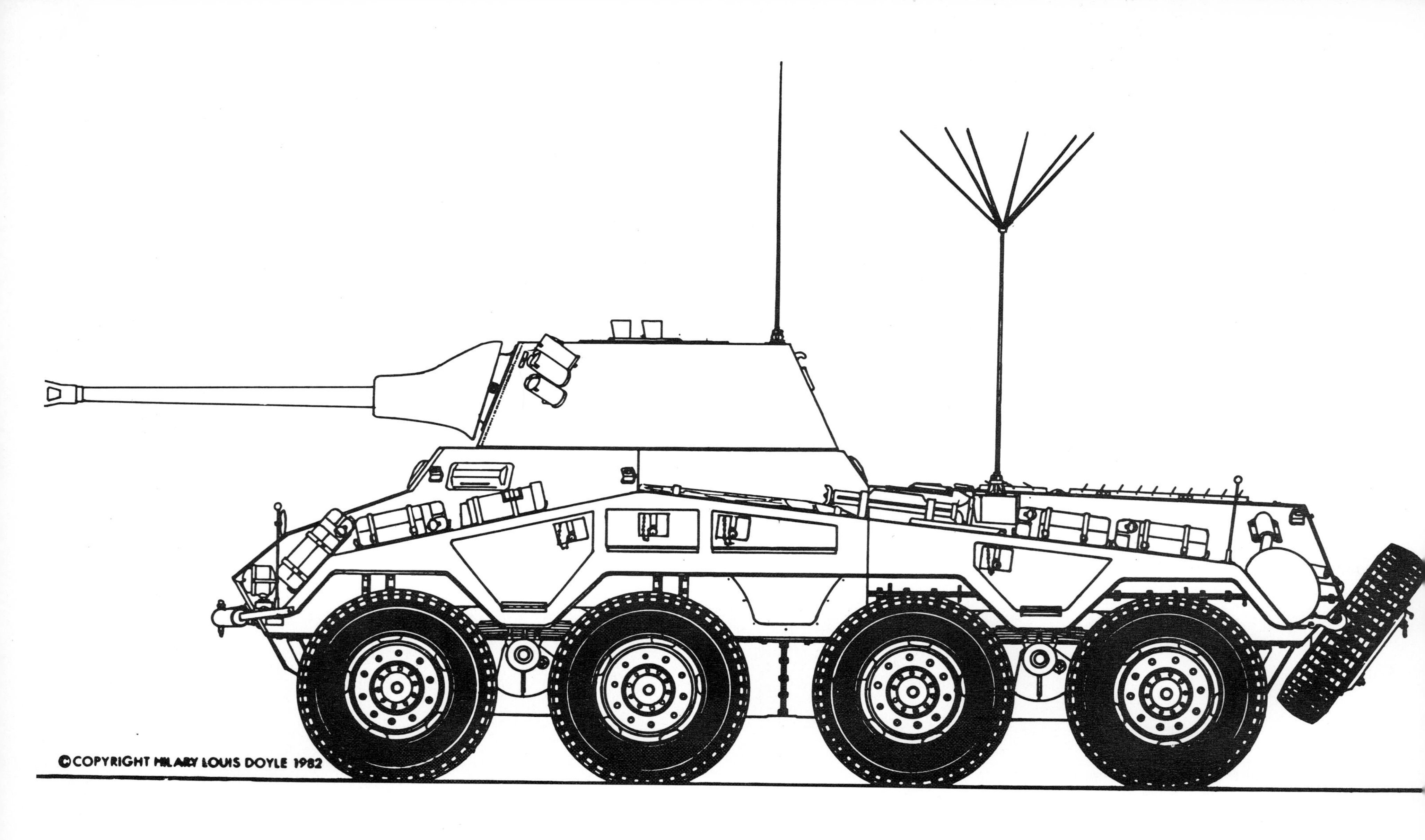

This drawing shows both the staff antenna (which all had) and the umbrella antenna (planned for longer ranges). Its 5 cm KwK 39/1 gun is mounted in a pig's-head mount. The crew consisted of four. 101 of these vehicles were built.

Above: One of the first Pumas to reach the troops, it still does not have the customary two- or three-color paint job of the time.

Upper right: On this one – as on the other – one can see the additional fuel canisters attached to the wheel covers.

Right: This view from the top shows the two lookouts for the turret crew. The ventilator opening can be seen behind them.

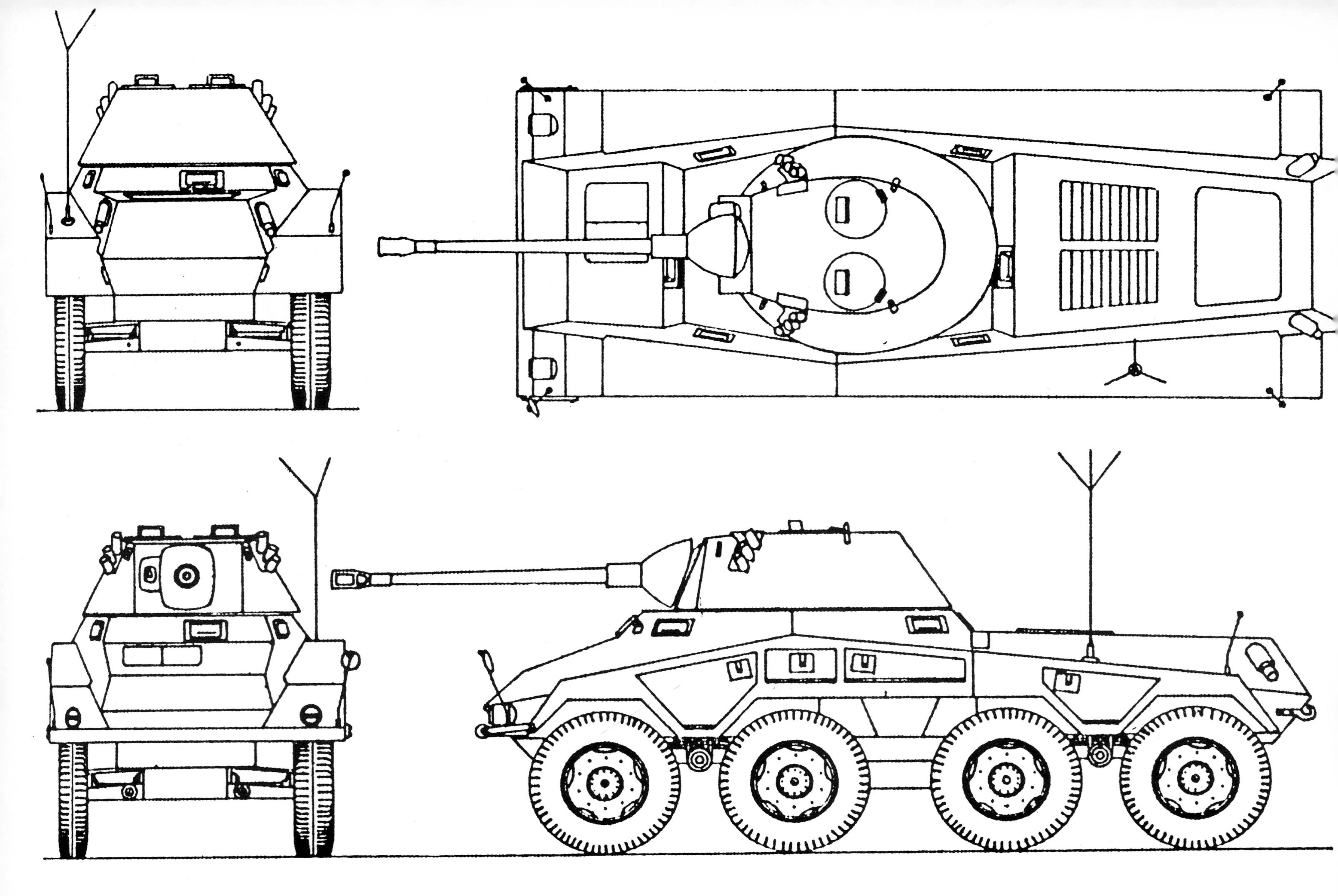

This four-way drawing shows the opening for the coaxial MG 42 machine gun.

The Puma already shown on page 9, seen from the left front. This Sd. Kfz. 234/2 shows irregular sprayed-on spots of olive green over a darker tan basic color.

The big, wide wheels are impressive. They made for better off-road capability than that of the heavy armored reconnaissance vehicles of the GS Series.

Here too, the vehicle seems to have been painted in only two colors.

Upper left and above: There are very few action photos of this interesting vehicle. Three men of its crew are sitting on top of it, outside the fighting compartment, as was customary when marching where there was no danger of enemy attack. Left: Technical service to the gun.

Left: Here too, the extension of the gun mount to hold the coaxial machine gun can be seen clearly. On either side of the turret are three launchers (operated from inside) for fog canisters.

Heavy Armored Reconnaissance Vehicle – Sd. Kfz. 234/3

Its fighting compartment was open at the top, and it was armed with the familiar – if also outmoded – 7.5 cm KwK 37 L/24, plus an MG 42 or 34. All in all, it strongly resembled the Sd.Kfz. 233, from which it can be told almost exclusively by the full-length wheel covers and the larger wheels. This vehicle, like the 233 before it, bore the designation "support vehicle."

Its crew consisted of four men. Along with the customary radio set "a", it also had a speaking tube for communication within the vehicle. Its frontal armor was 30-40 mm thick. It could carry 50 rounds of ammunition, and its gun could swing 12 degrees to either side.

A total of 88 were built – all of which managed to reach the troops. It was to be replaced by the 234/4 heavy reconnaissance vehicle, since the latter could also shoot down tanks with its longer 7.5 cm gun.

The open fighting compartment and the mounting of the 7.5 cm KwK L/24 can be seen clearly here. Everything strongly resembles the Sd. Kfz. 233 of the GS Series.

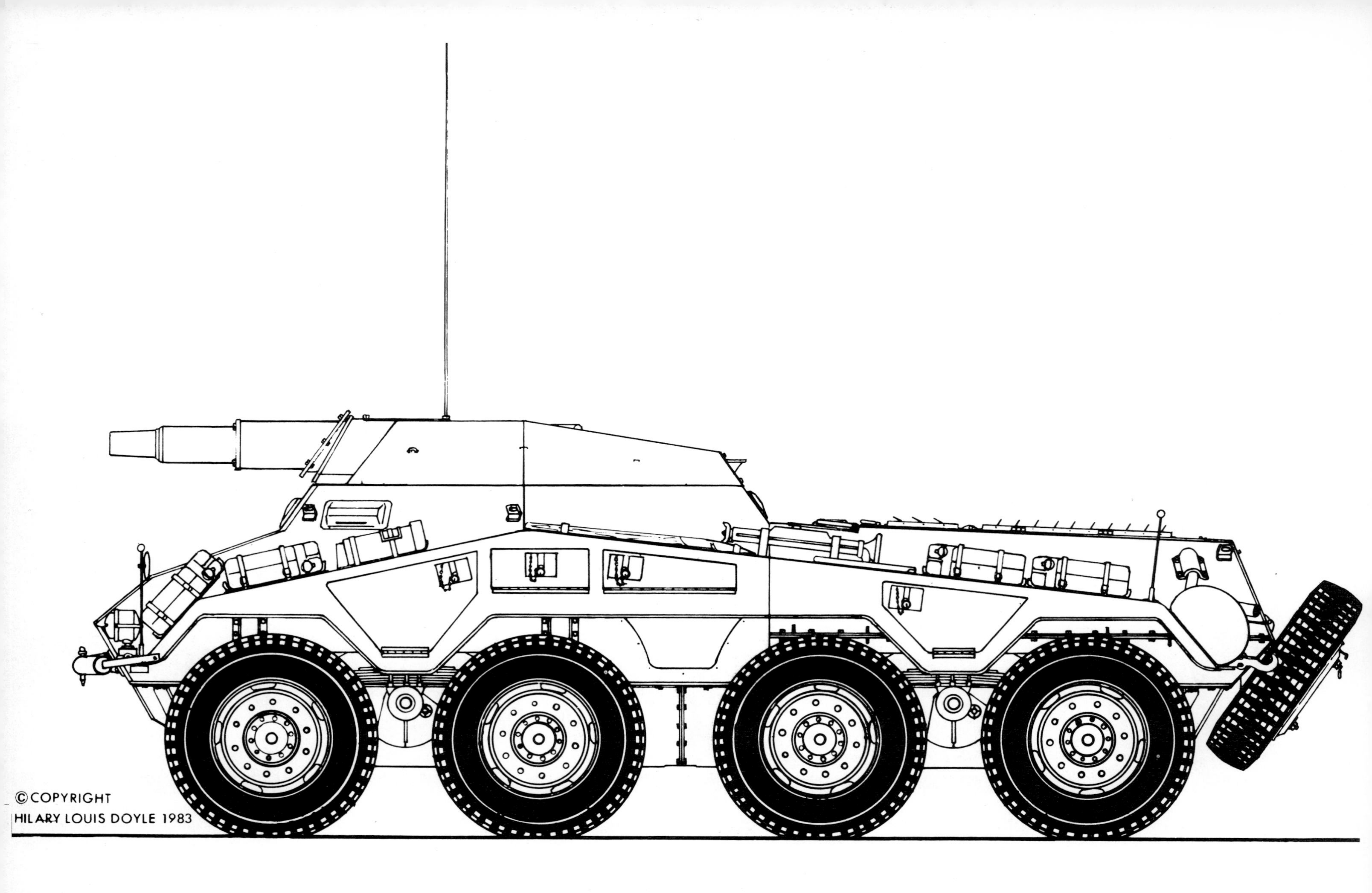

This drawing shows the attachment of the spare wheel at the rear. Here too, the spare fuel canisters can be seen at both ends of the wheel covers, along with the spade and jack.

This 234/3 does not yet have the usual 1944 paint job.

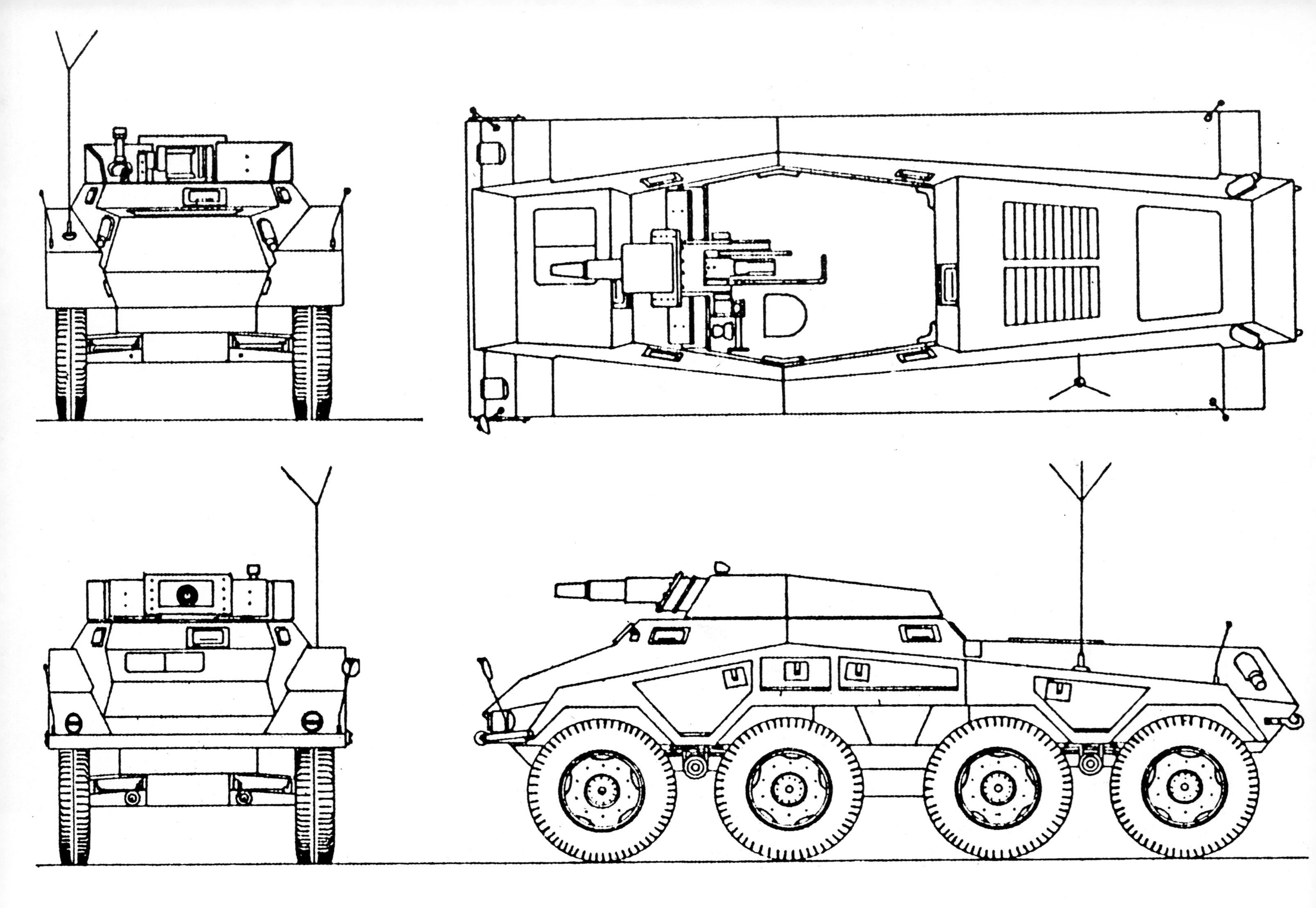

With its 11.5 tons of weight and 6 meters of length, it was not a small vehicle.

Left and above: This and the next two photos all show the same heavy armored reconnaissance vehicle, Sd. Kfz. 234/3. It was photographed at the U.S. Armored Troop School. It was planned as a "support vehicle" of the 234/1. Every armored reconnaissance unit had six of this or the 234/4 united in one platoon.

WH-1751008

Heavy Armored Reconnaissance Vehicle – Sd. Kfz. 234/4

At the end of 1944, Hitler ordered that the 7.5 cm "Stummel" gun used on the heavy armored reconnaissance vehicle, Sd. Kfz. 234/3, be replaced, on account of its strongly curved ballistic trajectory and its lack of power, by the successful 7.5 cm Pak 40. It was mounted on a pivot in the center of the fighting compartment, almost unchanged other than being removed from its previous mount. In this way, though, a risky design came into being – especially in view of the unfavorable recoil effects on the wheels and chassis. In addition to the longer gun, it had – as did the 234/3 already – an additional machine gun (MG 34 or 42). Along with the relatively meager twelve shells for the big gun, it carried 1950 rounds for the machine gun.

This vehicle was given the designation Sd. Kfz. 234/4, since the previously built 234/3 remained in use and a differentiation became necessary. By the end of the war, 89 of them had been built – but most of them did not reach the troops.

Plans for a firmer combination of vehicle and gun – utilizing a newer 7.5 cm AK B84 – could not be carried out.

The long (L/46) gun lengthened the vehicle to 6.68 meters.

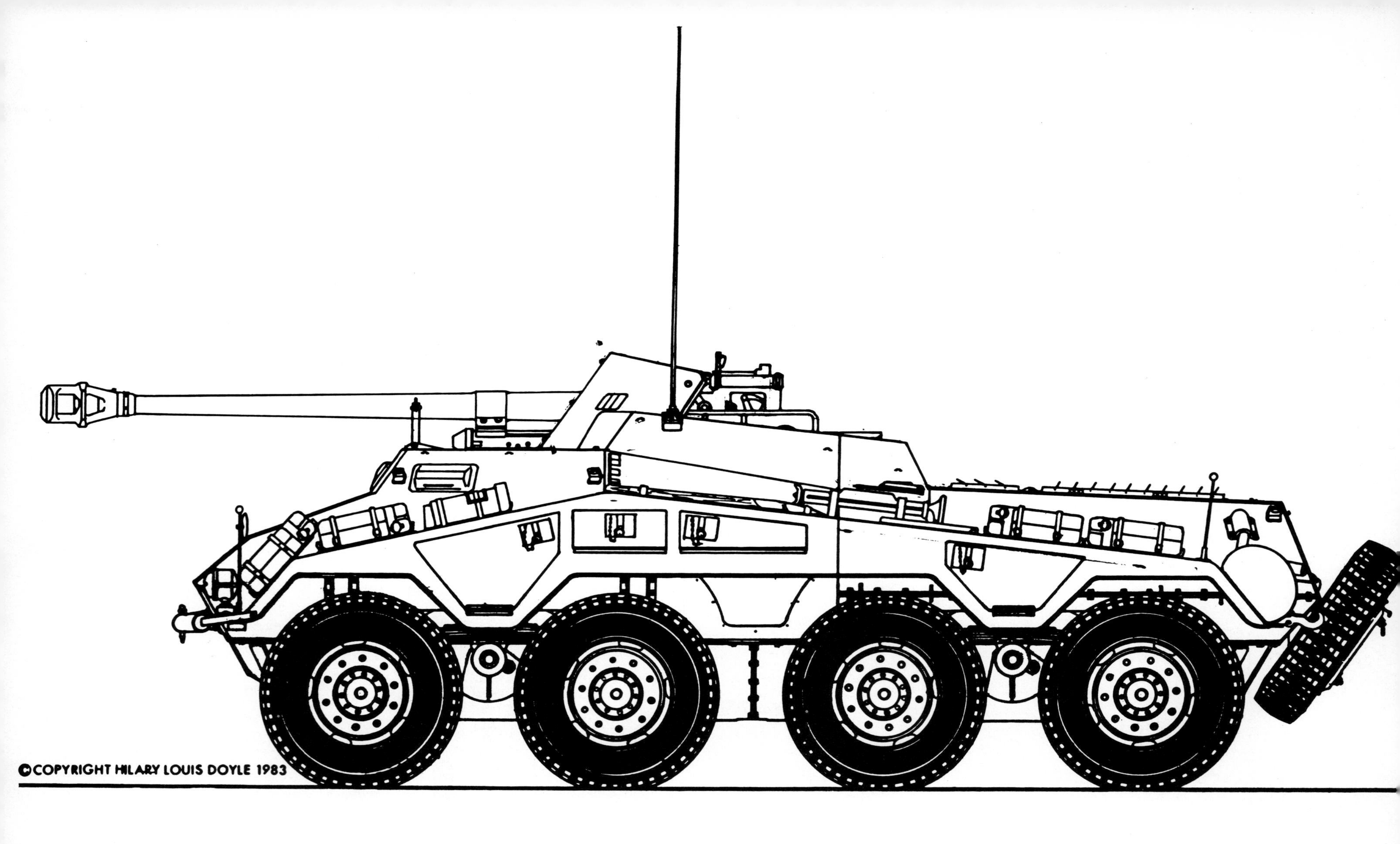

With its long, large-caliber gun, there was an unfavorable effect of recoil on the wheels and chassis.

Here we see the multicolored camouflage paint customary by the end of the war.

This and the facing photo show a Sd. Kfz. 234/4 restored at the Bundeswehr's Combat Troop School 2. The vehicle is presently at the Panzer Museum in Munster. It bears the symbol (a white steel helmet) of the "Gross-deutschland" Panzergrenadier Division. The symbol below for "reconnaissance" was not customary for GD at the end of the year.

WH-923350

Above: A look into the fighting compartment.

Upper left: One of the few action photos.

Opposite page: Another photo of the restored vehicle at Combat Troop School 2.

Left: This 234/4 is in an American museum. It lacks the wheel covers with the built-in storage chests. But one can still recognize the shot-deflecting shape (which still resembles the "coffin shape" of the 1932 six-wheeler!).

WH-923350

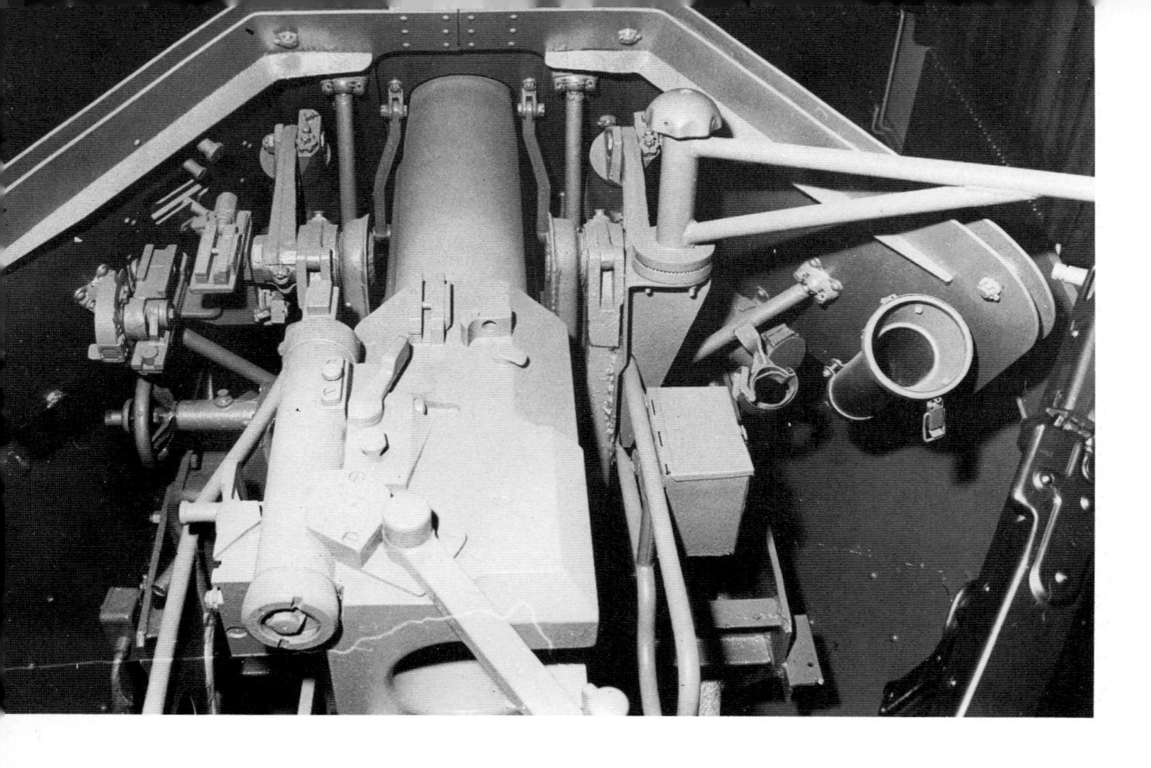

These photos show details of the gun mount and the aiming mechanism.

All the pictures on these pages were taken of the vehicle at the Panzer Museum in Munster. The photo above on this page shows the right side, that on the right shows the left side. The plate with a round hole is the "deflector" to protect the gunner, who sat on the left side of the gun.

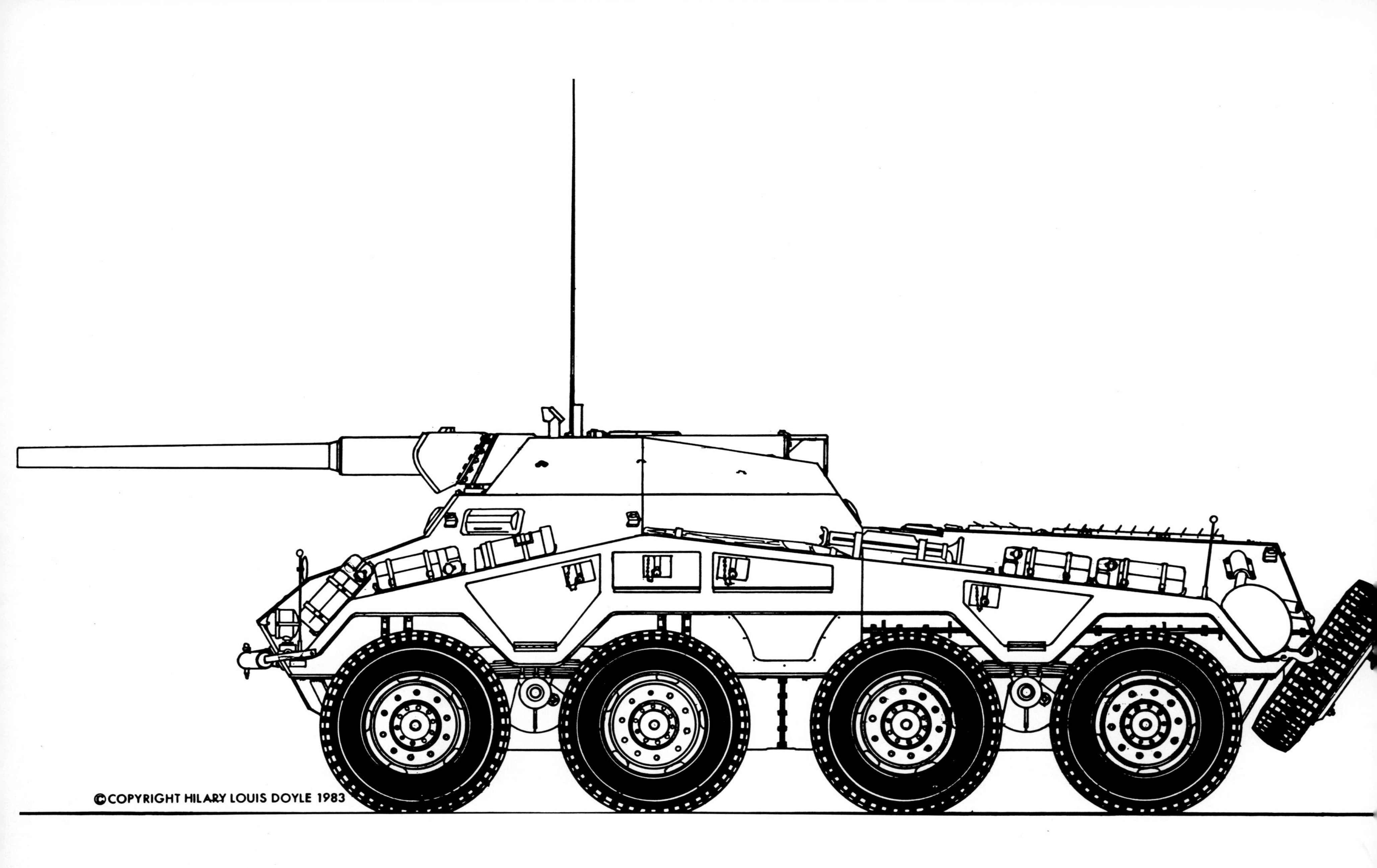

This drawing shows a planned further development with a new, even longer 7.5 cm KwK gun. It was designated Sd. Kfz. 234/4 with AK 7 B. But it was never built, not even as a prototype.

Wheeled/Tracked Vehicles built by Saurer (Vienna)

Beginning in 1930, the Austrian Saurer Works AG of Vienna developed a wheeled/tracked (R/R) chassis in developmental stages 1 to 7. After the inclusion of Austria in the Reich, this factory was given a contract by the Army Weapons Office to develop it into a reconnaissance vehicle that was safe from handguns and shrapnel. It ultimately ran under the designation of Panzerspähwagen RK (Rad/Kette)-9. Its armament consisted of only one machine gun in a rotating turret, which had been developed by the Daimler-Benz Works in Berlin in collaboration with the Schichau Works in Elbing. Only two prototypes of it were produced, in 1942. It was never put into production.

Meanwhile, though, in 1940 there had been a series of 140 "medium armored observation vehicles" based on the RR-7 development. It was given the designation of Sd. Kfz. 254 and used by the motorized artillery units of the Army. The change from wheels to tracks, or tracks to wheels, could be made while moving slowly, without stopping.

Below: One of the prototypes with open fighting compartment. The engine was a Saurer OKD (Diesel) producing 100 HP. Its top speed on tracks was 30 kph.

Below: The second prototype – now with a rotating turret and armed with one heavy (EW 141) and one light (MG 34) machine gun. On wheels, either type could reach a top speed of 80 kph.

Its precise designation was “medium armored observation vehicle.”

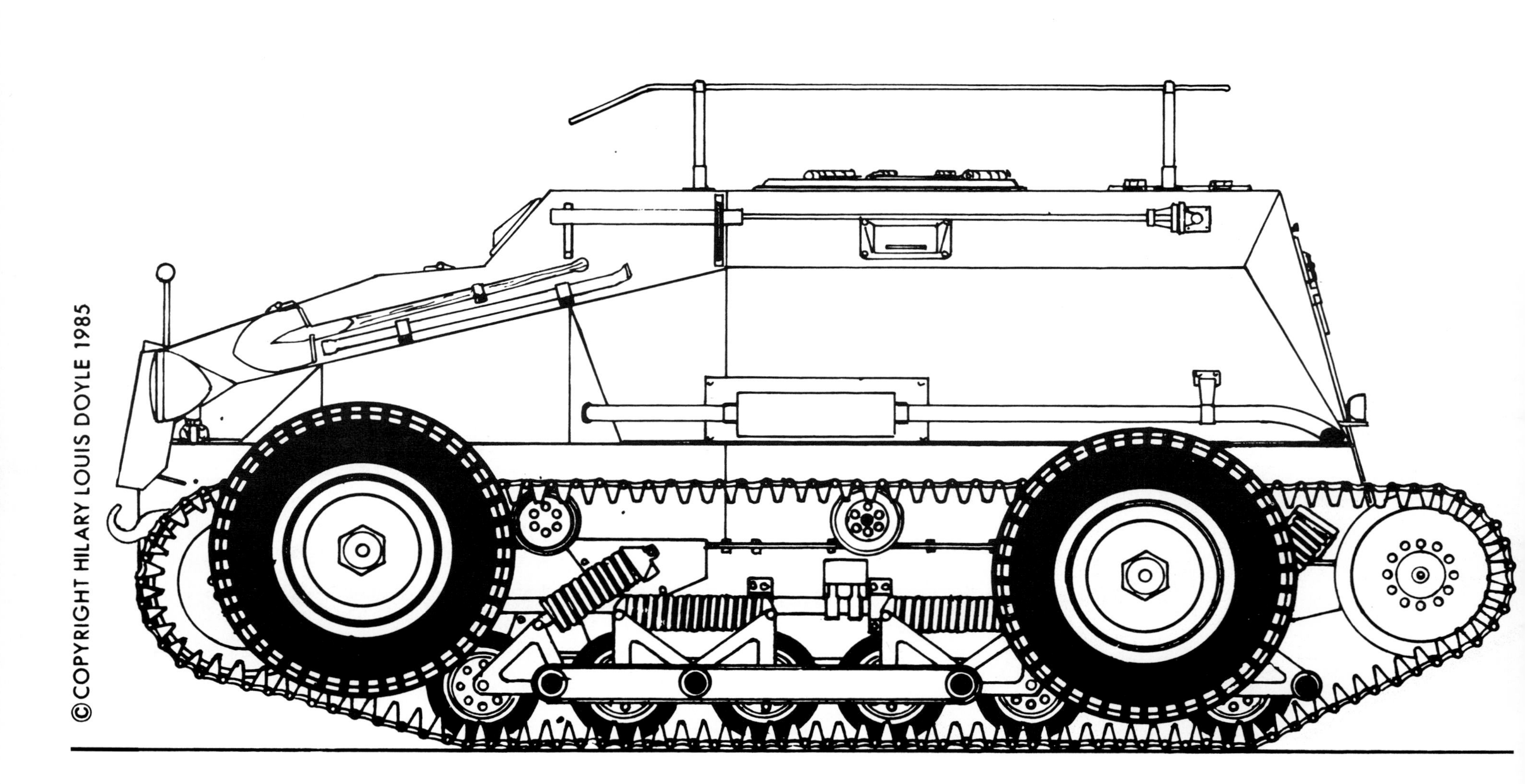

All Sd. Kfz. 254 were used as observation vehicles by the armored artillery units – as here, in Africa, as can be told by the symbol for the artillery on the nose.

Upper left: Here again is the prototype with the open fighting compartment. The picture clearly shows the vehicle's very wide track.

Above: Only 140 of these were built.

Left: This picture clearly shows the leading wheels for the tracks, set rather far to the rear, and the running gear, modified from that of the prototype.

A medium armored observation vehicle of Armored Artillery Regiment 78 of the 7th Panzer Division (note division symbol on the nose).

Above: Here a 254 of the 6th Panzer Division tows several trucks through deep sand before Leningrad in 1941. Surely this could not have been good for the vehicle, which was very damage-prone.

Both pages: Repairs in the desert sands. This was the Wehrmacht's only, but very striking, wheeled/tracked vehicle.

Captured Reconnaissance Vehicles

For the most part, these vehicles fell into the hands of the Wehrmacht during the Blitzkrieg actions of 1939 and 1940, and were put into service either after overhauling or directly upon capture. Further production in factories of occupied countries was also carried on. After the disarming of the Italian forces in 1943, a new impetus was provided.

With few exceptions, they were used only in small numbers by the German fighting forces, and then usually by the police or security forces in action against partisans in Russia or Yugoslavia.

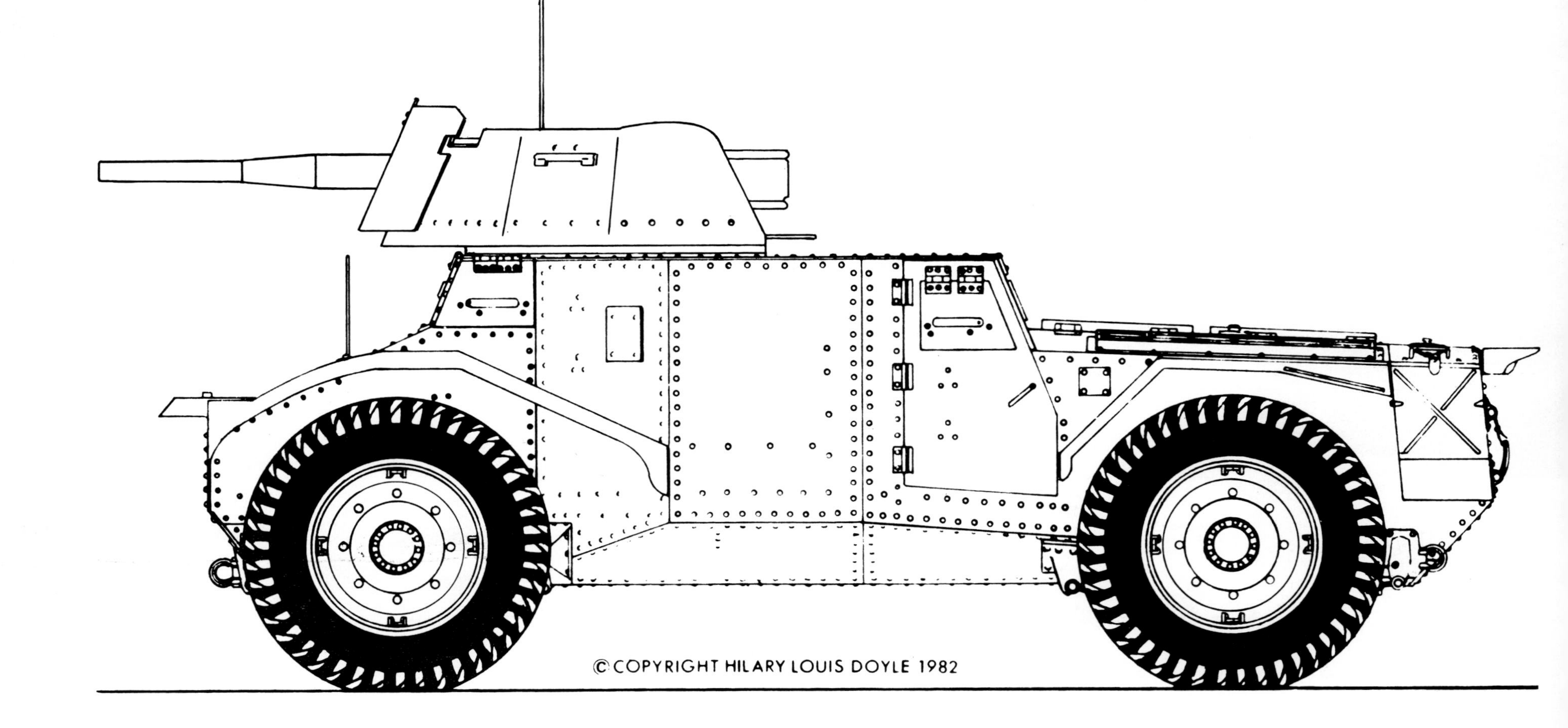

A police vehicle (with the police symbol next to the German cross) in action against partisans.

Upper left and above: British AEC armored command vehicles, which German staffs were happy to obtain.

Upper left: Rommel's command vehicle. On the front fender is painted the black-white-red corps standard of the Afrika-Korps, as well as the name "Moritz." Above is a vehicle of the 21st Panzer Division – note the division symbol on the fender.

A British Humber Mark II in German service – surely only as long as ammunition and spare parts lasted.

The Italian AB (Autoblinda: armored vehicle) 41 201(i).

It was taken over in large numbers from the Italian forces by the Wehrmacht in Italy and the Balkans after the disarmament of Italy (September 1943). A few were also used by the German forces in North Africa.

It was armed with a 2 cm M 35(i) L/65 gun and two 8 mm MG 38(i) machine guns – one of them for firing to the rear. This reconnaissance vehicle was built by the firm of SPA in Turin.

Afterword

The ARK Series – also in comparison to the enemies of the time – represented the technical high point of the construction of armored wheeled vehicles. When they were equipped with stronger weapons – thus following a trend of the later war years – this amounted to a wrong direction for the reconnaissance troops. Independent of the problems caused to the wheels by strong recoil, a heavy armament contradicts the basic function (seeing, hearing, reporting) of every reconnaissance troop. In this way they came into the danger – as was proved in the war – of having to function as combat troops (for which they were not trained) in crisis situations.

Heavy armament, large vehicles, and particularly those with tracks, do not belong in reconnaissance. Small, quiet, fast and nimble vehicles with good radio connections and great ability to avoid combat are called for. From this standpoint, the ARK Series, despite its technical excellence, especially its being armed with guns up to the 7.5 cm Pak L/48, was a wrong path.

Right: Medium armored observation vehicles of Armored Artillery Regiment 76 of the 6th Panzer Division at a training camp.

Technical Data

Heavy Armored Reconnaissance Vehicle ARK Series

Motor	Tatra Diesel with fuel injection
Cylinders	V 12
Displacement	14,825 cc
Power	210 HP at 2200 rom
Compression	1 : 16.5
Valves	Dropped
Cooling	Air; blower
Battery	Two 12-volt 120 Ah
Drive train	Rear engine, 8-wheel drive
Transmission	3 speeds + pre-selector
Axle ratio	5.70
Suspension	All on double transverse links
Foot brakes	Compressed air (Knorr)
Foot brakes affect	8 wheels
Hand brakes affect	8 wheels
Wheelbase	1300 + 1400 + 1300 mm
Track	All 1945 mm
Dimensions	6000 x 2330 x 2100 mm Sd. Kfz. 234/2 height 2380 mm Sd. Kfz. 234/4 height 2350 mm
Tires	270-20 (bullet-proof)
Ground clearance	350 mm
Wading ability	1200 mm
Turning circle	14.9 meters
Gross weight	11,700 kg
Top speed	90 kph
Fuel consumption	Road 40, off-road 60 l Diesel/100 km
Fuel capacity	240 or 360 liters
Range	Road 600/900, off-road 400/600 km
Armor	Front 30, sides 8, rear 14.5 mm
Crew	4 men
Armament	Sd. Kfz. 234/1: 2 cm KwK + MG Sd. Kfz. 234/2: 5 cm KwK + MG Sd. Kfz. 234/3: 7.5 StuK + MG Sd. Kfz. 234/4: 7.4 cm Pak

Medium Armored Observation Vehicle Saurer

Motor	Saurer "CRDv"
Cylinders	4 in-line
Displacement	5320 cc
Compression	16 : 1
Engine speed	2000 rpm
Power	70 HP
Power to weight	19 HP/ton
Valves	Dropped
Battery	4.6-volt, 87.5 Ah
Fuel system	Pump
Cooling	Water
Clutch	Single-plate dry
Top speed	60 kph
Range	Wheels/road 500, tracks/off-road 90 km
Turning circle	Wheels 15, tracks 8.5 meters
Suspension	Tracks; coil springs; wheels; torsion bars
Brake type	Inside drum
Foot brakes affect	4 wheels
Hand brakes affect	Gearbox
Wheel type	Sheet-steel disc
Tire size	8.25-20
Front/rear track	2000/1800 mm, tracks 1260 mm
Wheelbase	2400 mm
Track length	2390 mm
Track width	240 mm
Ground clearance	Wheels 300, tracks 240 mm
Overall length	4500 mm
Overall width	2470 (1960) mm
Overall height	2020 (2200) mm
Seats	4 (6)
Fuel consumption	Wheels/road 17, tracks/off-road 80 liter per 100 km
Fuel capacity	72 liters
Armor: front	14.5 mm
sides & rear	10 mm
Performance:	
Climbing ability	42 degrees, 900 mm
Wading ability	Wheels 900, tracks 750 mm
Spanning ability	2000 mm

Armored Reconnaissance Vehicle Panhard 178

Motor	Panhard "SS"
Cylinders	4 in-line
Displacement	6330 cc
Compression	5.8 : 1
Engine speed	2000 rpm
Maximum power	105 HP
Power to weight	12.6 HP/ton
Generator	12-volt
Batteries	Two 12-volt, 146 Ah
Fuel system	Pump
Cooling	Water
Clutch	Wet plates
Drive wheels	All
Top speed	72.6 kph
Range	350 km
Front axle	Rigid
Turning circle	8.0 meters
Suspension	Longitudinal semi-elliptic
Brake type	Disc
Foot brake affects	4 wheels
Hand brake affects	4 wheels
Wheel type	Sheet-steel disc
Tire size	42 x 9
Track	1737 mm
Wheelbase	3120 mm
Ground clearance	260 mm
Overall length	5140 mm
Overall width	2010 mm
Overall height	1651 mm, with APX 3 turret
Gross weight	8300 kp
Seats	4
Fuel consumption	Road 33, off-road 90 l per 100 km
Fuel capacity	125 liters
Armor plate	7 to 20 mm
Climbing ability	22 degrees
Wading ability	800 mm
Armament	One 25 mm KwK (150), 3 light MG (3750)

Also available from Schiffer Military History

OPERATION BARBAROSSA
OSTFRONT 1944
ROMMEL IN THE DESERT
PANZERKORPS GROßDEUTSCHLAND
THE HEAVY FLAK GUNS
THE WAFFEN SS
THE 1st PANZER DIVISION
THE GERMAN INFANTRY HANDBOOK